CONCEPTS AND TECHNIQUES IN MODERN GEOGRAPHY No. 3

UNDERSTANDING CANONICAL CORRELATION ANALYSIS

by

David Clark
(Lanchester Polytechnic)

CONTENTS

		Page
I	INTRODUCTION	
(i)	Types of correlation analysis	3
(ii)	Prerequisites	
(iii)	Applications in geography	
II	CONCEPTUAL OVERVIEW	
(i)	Data inputs	
(ii)	The canonical problem	7
(iii)	Canonical correlations and weights	8
(iv)	The canonical scores	9
(v)	Results and interpretation	12
III	THE MATHEMATICAL MODEL	
(i)	The partitioned intercorrelation matrix	12
(ii)	The canonical equation	14
(iii)	The significance of the roots	17
(iv)	The canonical vectors	17
(v)	The canonical scores	17
IV	A SIMPLE WORKED EXAMPLE	
(i)	The raw data	18
(ii)	Calculate means and standard deviations	18
(iii)	Standardise the raw data	18
(iv)	Calculate and partition the correlation matrix	19
(v)	Calculate the latent roots	19
(vi)	Test the significance of the roots	20
(vii)	Calculate the canonical weights for vector B_1	21
(viii)	Calculate the canonical weights for vector A_1	22

(ix)	Calculate the canonical scores	22
(x)	The complete results	23

V   A GEOGRAPHICAL APPLICATION

(i)	Theory and data	24
(ii)	Results	25
(iii)	Interpretation	26
(iv)	Evaluation	28

VI   PROBLEMS AND POSSIBILITIES

(i)	The characteristics of the input data	29
(ii)	Data formats	31
(iii)	N-modal canonical analysis	31
(iv)	Interpreting the results	31

VII   CONCLUSION   33

UNDERSTANDING CANONICAL CORRELATION ANALYSIS

"It is essential to arm yourself with a large bore computer loaded with a program of considerable weight: the canonical correlation is reputed to be dangerous when wounded." (Gould, 1968)

I INTRODUCTION

Geography is commonly defined as the scientific study of spatial systems. As the occurrence of most human and physical phenomena varies from place to place, these systems are typically highly complex, though the existence of basic principles and relationships in the organisation of space (for example, distance decay - see CATMOG 2), means that many of these systems have a great deal in common. It follows that a major requirement in geographical study is the measurement of the associations between complex spatial patterns. Such a set of measures is provided by canonical correlation analysis.

(i) Types of correlation analysis

Canonical analysis is one of a family of correlation techniques. Most students will be familiar with the simplest index of correlation, namely Pearson's product moment correlation coefficient, which specifies the association between two variables, but other, higher order correlation procedures, are required to study multivariate associations. Together with the product moment coefficient, these more sophisticated forms of correlation permit the analysis of the three basic types of relationship among geographical data:

a) Product Moment Correlation measures the relationship between the observed values of *two* variables.
b) Multiple Regression Analysis measures the relationships between the observed values of *one* variable, and the observed values of *a set* of variables.
c) Canonical Correlation Analysis measures the relationships between the observed values of *two sets* of variables.

Canonical correlation is the most general form of correlation. Multiple regression analysis is a more specific case in which *one* of the sets of data contains only one variable, while product moment correlation is the most specific case in that *both* sets of data contain only one variable. But canonical correlation analysis is *not* related to factor/principal components analysis despite certain conceptual and terminological similarities. Canonical correlation analysis is used to investigate the intercorrelation *between two sets* of variables, whereas factor/principal components analysis identifies the patterns of relationship *within one set* of data. This difference between the two techniques is fundamental and requires special emphasis; it means that the geographical problems studied by canonical correlation analysis cannot also be studied by factor/principal components analysis.

(ii)   Prerequisites

In view of their close association, a basic understanding of multiple regression analysis is useful, but not crucial, to the understanding of canonical correlation analysis. The essential perequisites are four in number; an '0' level knowledge of maths, a fair grasp of simple correlation and linear regression, a familiarity with the chi squared test of significance, and a working knowledge of simple matrix algebra, particularly the techniques of matrix multiplication and inversion. The second and third topics will be familiar to most first year undergraduates since they are covered in detail by the standard introductory texts dealing with statistics in geography (Gregory, 1963; Theakstone & Harrison, 1970). Elementary matrix algebra may be less well known, but the techniques are fairly straightforward since they merely involve simple arithmetic. Matrix methods for geographers are discussed by King (1969) and Cole & King (1968).

(iii)   Applications in geography

Despite the fact that geographers have made extensive use of multiple regression analysis, canonical correlation analysis has received comparatively little attention. In *Statistical Analysis in Geography*, one of the first texts of its type to be aimed at the final year undergraduate, and postgraduate reader, L.J. King (1969) identifies canonical correlation as one of the most promising developments in quantitative spatial analysis, but quotes only two applications of the technique, those of Berry (1966), and Gauthier (1968). This restricted usage can partly be explained by the limited availability of computer programs, and the fact that King's is the first geography text to detail the procedures. But there must be many aspiring analysts who have progressed no further than volume III of the statisticians 'bible', Kendall and Stuart's *The Advanced Theory of Statistics* (1966). Their considered opinions that 'the literature of the subject has few examples of useful applications', or worse still, that 'the results of canonical correlation are even more difficult to interpret than those of components analysis' (p. 305) carry all the weight of the ultimate deterrent!

Geographical applications of canonical correlation analysis remain limited in number, are largely experimental, and in any case, are very much a feature of the last decade (Table 1). They begin with the observed values of two sets of variables relating to the same set of areas, and a theory or hypothesis that suggests that the two are interrelated (figure 1). We may, for example, wish to measure the association between the growth of urban centres and their accessibility within the urban system (Norcliffe, 1972), or the link between the retail trading performance of places and the range of shops and businesses they possess (Andrews, 1973). Other examples include the relationship between housing morphology and social class (Openshaw, 1969), and the connection between criminal deviance and socio-economic character (Gittus and Stephens, 1974). In all of these cases, the overriding concern is with the structural relationship between the two sets of data *as a whole*, rather than the associations between individual variables. It is for this reason that these researchers use canonical analysis in preference to other more simple forms of correlation.

But despite the growing range of applications, some basic difficulties remain. Canonical correlation is *not* the easiest of techniques to follow,

Table 1. Geographical Studies Using Canonical Correlation Analysis

Study	Variables				Observations	
	Criteria	p	Predictors	q	Areas	N
Andrews, 1973	Urban structure	4	Trading performance	6	British towns	155
Andrews, 1973	Urban structure	4	Retail equipment	6	British towns	155
Berry, 1966	Spatial behaviour	5	Spatial structure	7	Pairs of places in India	1260
Clark, 1973a	Urban activities	5	Urban linkages	8	Welsh telephone charge groups	72
Clark, 1973b	Flow origins 1958	8	Flow origins 1968	8	Welsh telephone charge groups	72
Clark, 1973b	Flow destinations 1958	8	Flow destinations 1968	8	Welsh telephone charge groups	72
Gauthier, 1968	Economic development	3	Highway accessibility	8	Brazilian towns	123
Gittus and Stephens, 1974	Crime	20	Socio-econ character	18	Seattle census tracts	93
Norcliffe, 1972	Functional structure	2	Demographic change	2	Irish towns	26
Openshaw, 1969	Urban building fabric	9	Urban social structure	9	South Shields E.D.s	49
Ray, 1971	Economic variables	20	Cultural variables	17	Canadian census counties	184

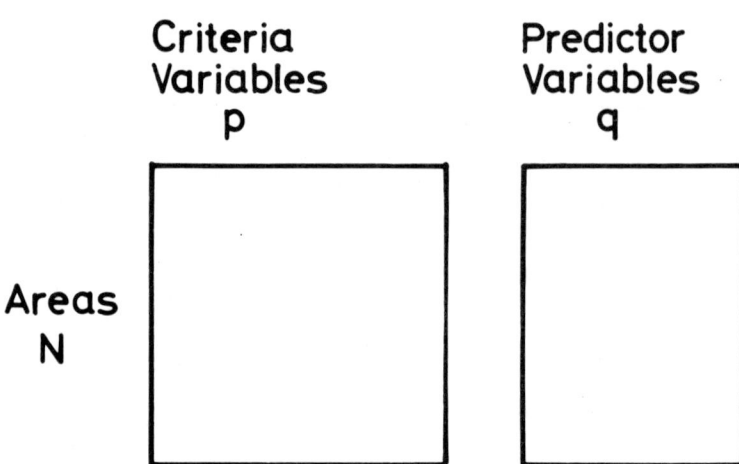

Fig. 1 The starting point for canonical correlation analysis in geography: two sets of variables relating to the same set of areas.

though the problems of comprehension are conceptual rather than mathematical. Unlike multiple regression and principal components analysis, we cannot provide a graphic device to illustrate even the simplest form. For with canonical correlation analysis we are dealing with *two sets* of data. Even the most elementary example must, therefore, have at least two variables on each side and so we require 2 + 2 = 4 dimensions. Tied as we are, however, to a three dimensional world, a true understanding of the technique in the conventional cognitive/visual sense of the term, is beyond our grasp. In view of this difficulty, it was thought sensible to adopt a wide range of approaches. Rather than rely upon any single treatment, this monograph works through canonical correlation analysis in four different ways, conceptually, mathematically, numerically and empirically. Only then do we turn to the specific problems and potentials that arise when the technique is applied to spatial data.

II CONCEPTUAL OVERVIEW

(i) <u>Data inputs</u>

We introduce canonical correlation analysis by considering a hypothetical example in purely conceptual terms. Let us take a situation in which we

Table 2 Hypothetical Employment Data for Simple Canonical Correlation Analysis

		YEAR ONE						YEAR TWO				
		Criteria Employment			Variables			Predictors Employment			Variables	
		1	2	3	4	5	6	7	1	2	3	4	5
A	Zone 1	09	81	100	1000	0.1	21	53	97	0.7	1001	27	303
R	Zone 2	11	09	700	7000	0.7	23	65	12	0.1	1807	31	411
E	Zone 3	07	04	605	6035	1.9	24	88	01	0.0	0991	39	120
A	Zone 4	16	81	357	1972	4.9	22	99	99	0.0	0301	43	140
S	Zone 5	08	77	087	3315	8.9	26	49	88	0.1	5119	55	199
	Zone 6	11	69	420	0497	8.7	25	87	87	0.5	8007	27	613
	Zone 7	09	07	199	4414	7.6	24	66	03	3.2	5000	31	810
	Zone 8	07	10	148	3937	3.3	25	77	14	6.1	0098	33	098
	Zone 9	06	18	152	6163	3.7	22	81	17	0.0	9001	18	120
	Zone 10	01	01	433	0999	0.1	21	78	02	0.1	6330	63	140

are interested in measuring the changes over time of economic activity in a particular region. This problem can be operationalised by abstracting from the census, data on employment in a number of occupational categories for all the areas in the region, for two different dates.

Two matrices of hypothetical data, which might form the basis of such a study, are shown in Table 2. Note

a) The size of the matrices. There is no requirement in canonical analysis that there must be the same number of variables (columns) in each matrix, though there must be the same number of areas (rows). (There must of course be more than one variable in each set otherwise we would be dealing with multiple regression analysis).

b) The order of the matrices. Neither set of data is given priority in the analysis so it does not matter which we term the criteria and which the predictors. Unlike simple linear regression there is no concept of a 'dependent' set or an 'independent' set. But in practice the smaller set is always taken second as this simplifies the calculation enormously, as we shall see in section III, (ii).

(ii) The canonical problem

The individual variables in our analysis clearly interrelate with each other to define complex patterns of employment. Though there are several distinct categories of employment, the variables index the major occupational types such as "extractive industry" or "office activity" etc. For any regional economy to function, *some* employment is necessary in each of these major categories, but over time the balance may shift, both structurally (as, for example, in the general shift of employment from the manufacturing, to the service sector), and spatially, (as in the industrial decline in the north, and the expansion of the south of England economy). It is possible that these changes will be so small as to be negligible, in which case *all* the structural and spatial patterns of employment observed at the first date will be observed at the second. Conversely, there may be a total transformation in industrial activity so that *none* of the patterns originally present is recognisable at the later date. But the most likely situation is that there is *some* stability and *some* change. Certain patterns of employment will be common to both years, but others will only be present in one. Canonical correlation analysis identifies these common patterns. It tells us how important they are, and which types of industry, and which areas, are involved in them.

Clearly the problem is one of assessing the extent to which the variables intercorrelate to define the same general patterns of employment in one year as they do in the other. In a simple two-variable case the nature of the similarity between the variables can be expressed by plotting the data on the familiar two-dimensional graph. But in our example we are dealing with many variables and so we have to try and imagine patterns of intercorrelation of variables in n-dimensional statistical space. We have in fact two clusters of vectors, one defined by the intercorrelations of the variables for the first year p, the other composed of the intercorrelations of the variables for the second year q. In essence, the task is to assess the amount of overlap - to what extent do the vectors occupy the same positions in p-variable space as they do in q-variable space? Note of course that as we are working in an n-dimensional context there will be several different ways in which the patterns will coincide. (Empirically, what we are saying is that manufacturing employment in year one may be substantially the same as in year two, service employment in year one may be the same as in year two, and so on). The number of overlaps will be the same as the number of variables in the smaller data set: we cannot relate five to seven in more than five ways!

(iii) <u>Canonical correlation and weights</u>

The solution is similar to linear regression in that we use axes to summarise the relationships in the data. In canonical analysis, we put axes through each of the two clusters of vectors so that they account for the maximum amount of variance within each set and, at the same time, the maximum amount of covariation between each set. In other words the first pair of axes identifies the most important pattern that is common to the two sets of data. Once the best positions are found for this first pair, further pairs of axes may be stuck through the vector clusters at right angles to the first set, and rotated until they account for the next best correlation between the sets, and so on, until all the possibilities are exhausted. As we have said, this occurs when there are as many pairs of axes as there are variables in the smaller variable set.

The process of putting axes through clusters of vectors in n-dimensional statistical space is known as factoring and is common in multivariate statistics. For conceptual purposes it is useful to return to the two variable case, for in simple linear regression, we fit a single axis (a regression line) through two vectors (in two dimensional space) according to a single criterion (that it minimises the total sum of squared deviations). Essentially the same feat can be performed in a multivariate context except that we fit our axes to n vectors in n-dimensional space. By applying weights to each of the variables it is possible to insert a straight line axis through each cluster of vectors according to certain specified criteria. In canonical correlation there are three interdependent criteria:

a) <u>Maximum covariance</u> Both axes must identify the same pattern in their respective sets of data.
b) <u>Maximum variance reduction</u> The axes must account for as much of the variance as possible. The first pair of vectors identifies the greatest area of overlap, the second pair identifies the second greatest area of overlap and so on.
c) <u>Orthogonality</u> Each axis must be uncorrelated with other axes already fitted to the data. (In statistical terms this means that the axes are put through the clusters of vectors at right angles to each other).

The axes identify, in decreasing order of importance, the 'canonical dimensions' or 'communal patterns' which are common to the two sets of data. The importance of these canonical overlaps is expressed by a latent root, λ, which has a range of 0 to +1.0. The first latent root represents the maximum possible correlation between any weighted linear combination of the criteria variables, and any weighted linear combination of the predictor variables. The second latent root represents the next highest possible correlation between any weighted linear combination of the criteria variables, and any weighted linear combination of the predictor variables, and so on. Bartlett's chi squared test can be used to show which of the roots is significant. As there are five common patterns in our data, there are five latent roots, the first three being highly significant (Table 3).

The axes are merely column vectors made up of weights, one weight for each variable. As we are dealing with two sets of data there will be a pair of column vectors for each common pattern, one for the criteria variables, and one for the predictor variables. In our example, there are five common patterns and so there are five pairs of vectors, each vector having seven weights corresponding to each of the seven criteria variables, and each B vector having five weights, one for each of the five predictor variables. The canonical weights are shown in Table 3.

(iv) The canonical scores

Each pair of canonical vectors defines a pattern of employment which is common to the two sets of data. For example, the weights on the first pair of vectors may well indicate that 'extractive industry' is an occupational dimension that is common to the two years; the second pair may denote 'light manufacturing industry' and so on. These dimensions are common to the two years because areas within the region have about the same amounts of employment in these activities in year one as in year two. All of the common patterns have both a structural and a spatial meaning.

The structural identity is conveyed by the canonical weights while the spatial patterns are conveyed by the final set of measures in canonical analysis, the canonical scores.

Table 3 Hypothetical Employment Data: Latent Roots and Canonical Weights.

COMMON PATTERNS

Separate patterns which are common to the two sets of data

A latent root: indicates the size of a common pattern

	1	2	3	4	5
λ	1.00	1.00	1.00	0.80	0.25
Bartlett's chi-squared criteria χ^2	141.06	95.91	49.82	4.82	0.74
Probability	0.00	0.00	0.00	0.77	0.86

The probability that the common pattern could have arisen by chance

		λ_1		λ_2		λ_3		λ_4		λ_5	
V		A_1	B_1	A_2	B_2	A_3	B_3	A_4	B_4	A_5	B_5
A	1	-0.53	0.73	-1.16	-0.70	0.02	0.17	-1.07	0.36	-0.36	-0.05
R	2	1.31	-0.48	0.91	-0.80	1.00	-0.13	0.72	0.67	0.20	-0.00
I	3	0.64	0.11	0.74	0.15	-0.50	-0.45	-0.56	0.96	0.48	-0.35
A	4	0.26	0.10	0.98	-0.13	0.65	-0.57	0.25	0.12	-0.74	0.84
B	5	0.67	0.13	0.63	-0.27	-1.38	-0.74	0.05	-0.78	-0.94	-0.05
L	6	-0.59		-0.80		0.61		0.20		0.84	
E	7	-0.11		0.21		0.65		0.68		-0.08	
S											

A weight: degree and direction of involvement of the fourth criteria variable in the first common pattern

A weight: degree and direction of involvement of the fifth predictor variable in the fourth common pattern

Table 4. Hypothetical Employment Data: Canonical Scores

| | | COMMON PATTERNS ||||||||||
| | | 1 || 2 || 3 || 4 || 5 ||
		S_{a_1}	S_{b_1}	S_{a_2}	S_{b_2}	S_{a_3}	S_{b_3}	S_{a_4}	S_{b_4}	S_{a_5}	S_{b_5}
A	ZONE 1	0.80	0.80	-0.71	-0.71	1.01	1.01	-0.58	-0.18	-0.40	0.60
R	ZONE 2	-0.28	0.28	0.71	0.71	0.01	0.01	-1.15	-2.32	-0.14	0.16
E	ZONE 3	-0.56	-0.56	1.11	1.11	0.72	0.72	-0.91	-0.05	0.54	0.63
A	ZONE 4	1.00	1.00	-0.51	-0.51	0.99	0.99	-0.33	-0.49	0.74	-0.87
S	ZONE 5	1.10	1.10	-0.34	-0.34	-0.40	-0.40	0.91	0.85	1.00	-0.08
	ZONE 6	1.10	1.10	-0.54	-0.54	0.92	0.92	0.29	0.21	-1.20	0.58
	ZONE 7	-0.84	-0.84	-0.69	-0.69	-1.81	-1.81	-0.95	-0.52	-0.56	-1.36
	ZONE 8	-1.19	-1.19	-1.49	-1.49	0.80	0.80	0.98	0.68	0.21	0.47
	ZONE 9	-0.18	-0.18	1.44	1.44	0.57	0.57	1.13	1.11	-1.69	-1.64
	ZONE 10	-0.20	-0.20	1.06	1.06	-1.07	-1.07	0.82	0.63	1.51	1.50

A score: degree and direction of involvement of the seventh observed value of the criteria variables in the first common pattern.	A score: degree and direction of involvement of the ninth observed value of the predictor variables in the fourth common pattern.

The canonical scores (known also as canonical variates) identify the associations of the observed values of each of the input variables in each of the common patterns. They are derived by applying the canonical weights to the standardised raw data. This yields a value for each area and so, in our example, with five common patterns there are five pairs of canonical scores, each vector having ten score values, one for each of the ten areas (Table 4). The correlation between the two sets of scores is equal to the square root of the corresponding latent root (that is, it is equal to the canonical correlation). Vectors of canonical scores have the property of being orthogonal to one another within the set of criteria variables and within the set of predictor variables.

(v) Results and interpretation

The canonical correlation analysis results in three sets of measures, a set of latent roots, a set of pairs of canonical vectors (or canonical weights) and a set of pairs of canonical scores (or canonical variates). These measures convey the following information:

a) Latent roots. (plus associated indices of statistical significance) indicate the *size* of the patterns which are common to the two sets of data.
b) Canonical weights indicate the involvement of each of the *variables* in each of the common patterns.
c) Canonical scores indicate the involvement of each of the *areas* in each of the common patterns.

The weights and scores can be interpreted so as to reveal the structural and the spatial identity of the common patterns defined by the analysis. For purposes of interpretation we consider only those canonical dimensions which Bartlett's chi squared test shows to be significant. Both weights and scores are interpreted by considering the magnitude of each value in relation to the direction of their signs.

III THE MATHEMATICAL MODEL

(i) The partitioned intercorrelation matrix

Having discussed canonical correlation analysis in purely conceptual terms, we can now turn to the mathematics of the technique by considering the major steps in our hypothetical employment example (figure 2). It is useful at this stage to recall that canonical analysis identifies the patterns of interrelationship between two multivariate sets of variables, and that these common or canonical patterns are identified by vectors or axes which are put through the variables according to certain specified constraints.

In mathematical terms, the problem is to find a linear combination (a straight line axis) of the p variables, and a linear combination of the q variables so that their overall correlation is maximised. The nature of this correlation depends upon the patterns of association *within* and *between* the two data sets. These relationships can be expressed by combining the two sets of data and by calculating product moment correlation coefficients for each pair of variables. In our employment example there were 7 criteria variables and 5 predictor variables so that the dimensions of the correlation matrix R are 12 x 12. This matrix can be partitioned so that

$$R = \begin{array}{|c|c|} \hline R_{11} & R_{12} \\ \hline R_{21} & R_{22} \\ \hline \end{array}$$

where
R_{11} is the matrix of intercorrelations among the p criteria variables
R_{22} is the matrix of intercorrelations among the q predictor variables
R_{12} is the matrix of intercorrelations of the p criteria with the q predictors
R_{21} is the transpose of R_{12}

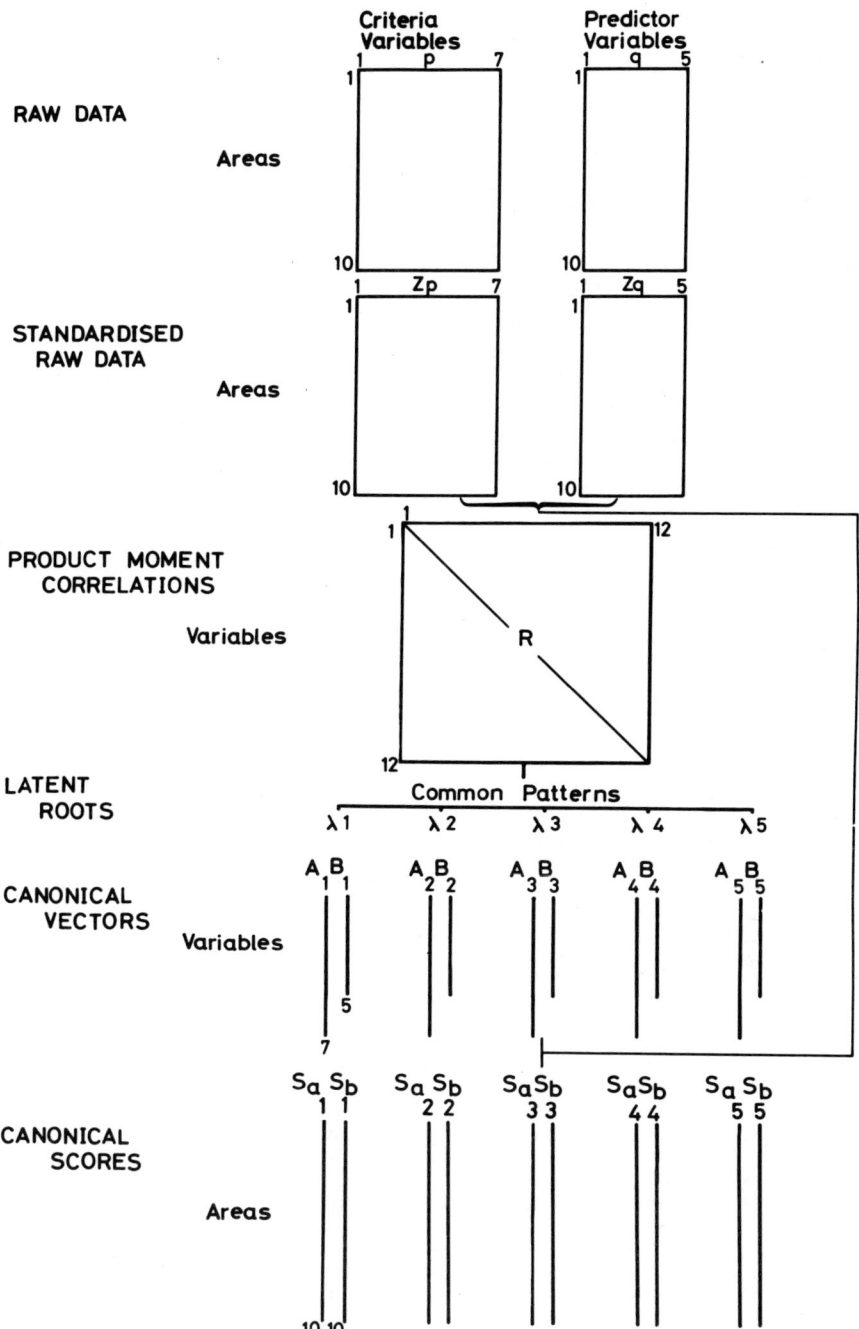

Fig. 2 Canonical correlation: the major steps in the analysis.

(ii) The canonical equation

The partitioned intercorrelation matrix expresses the basic patterns of association within and between the two sets of data. To find the common patterns, or canonical correlations, we first have to find the latent roots of the canonical equation:

$$(R_{22}^{-1} \cdot R_{21} \cdot R_{11}^{-1} \cdot R_{12} - \lambda I) = 0$$

where λ is the latent root and I is an identity matrix.

Few would deny that the appearance of such an expression in the geographical literature is unlikely to inspire the confidence of the undergraduate reader! And yet this statement lies at the heart of canonical analysis so we must try and see what it means. Fortunately, the formula involves two distinct sets of operations which can be discussed separately. In the first step, $R_{22}^{-1} \cdot R_{21} \cdot R_{11}^{-1} \cdot R_{12}$, we simply combine the partitions of the correlation matrix to produce a single product matrix which we can call M. We can regard the values in this matrix as expressing the ways in which the data interrelate or overlap. The size of the common patterns within the matrix is then identified by the latent roots, λ of the matrix. It is this task of extracting these roots that is performed by the second part of the canonical equation, $(M - \lambda I) = 0$.

a) The product matrix

The first part of the canonical equation, the creation of matrix M, involves a sequence of matrix multiplications and inversions. These operations might appear formidable but the arithmetic is in fact, fairly simple, as our worked example (section IV) shows. Far more difficult and far more important, is to try and visualise exactly what is happening in n-dimensional statistical space during these operations. For this purpose, Gould (1968) suggests that it may be useful to think through the geometry of a far more familiar algebraic statement:

$$R_{22}^{-1} \cdot R_{22} = I$$

in which we multiply a matrix by its inverse so as to produce an identity matrix.

In this simple case, if R_{22} is a matrix of correlation coefficients considered as the cosines between the vectors representing the variables in the q set, then R_{22}^{-1} is a linear operator that rotates the same vectors around so that they remain perfectly correlated with themselves but totally uncorrelated with each other. In other words, R_{22}^{-1} operating on R_{22} produces a completely orthogonal (uncorrelated) set of vectors.

But in canonical correlation, the linear operator R_{22}^{-1} operates not on sub-matrix R_{22}, which represents the correlations *within* the set of predictors, but on sub-matrix R_{21} which specifies the correlations *between* the predictors and the criteria. R_{22}^{-1}, the linear operator that orthogonalises the set of predictors, takes the criteria on one hand and the predictors on the other, and rotates them relative to each other. Similarly, in the second half of the formula, R_{11}^{-1}, which would orthogonalise the criteria vectors in R_{11}, operates upon R_{12} (that is, again, the matrix of correlations *between* the criteria and the predictors), so that the same angles change. In this way the two linear operators create a matrix M which expresses the overlap between the two sets of data. In the simplest of terms these operations are:-

$$R_{22}^{-1} \cdot R_{21} \cdot R_{11}^{-1} \cdot R_{12}$$

| Linear operator for the predictors | × | Correlation between the criteria and predictors | × | Linear operator for the criteria | × | Correlations between the criteria and predictors |

The sequence of matrix multiplications is:

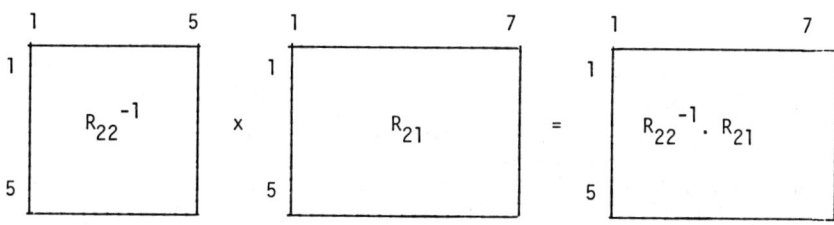

$$R = \begin{array}{c|cc} & \text{Criteria} & \text{Predictors} \\ \hline \text{Criteria} & R_{11}\ (7\times 7) & R_{12}\ (7\times 5) \\ \text{Predictors} & R_{21}\ (5\times 7) & R_{22}\ (5\times 5) \end{array}$$

First we have to evaluate $R_{22}^{-1} \cdot R_{21}$

$$\underset{5\times 5}{R_{22}^{-1}} \times \underset{5\times 7}{R_{21}} = \underset{5\times 7}{R_{22}^{-1}\cdot R_{21}}$$

Matrix of cosines between the predictors and the criteria after being rotated by R_{22}^{-1}

15

Then, $R_{11}^{-1} \cdot R_{12}$

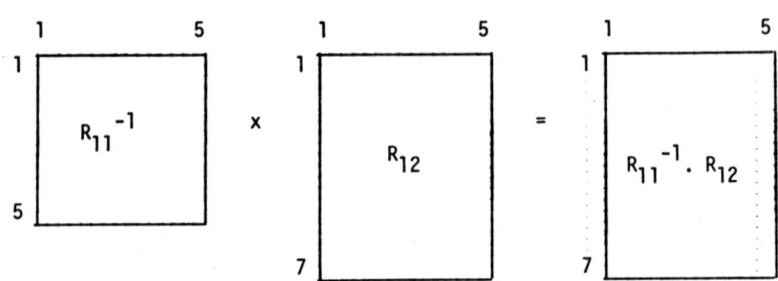

Matrix of cosines between the criteria and the predictors after being rotated by R_{11}^{-1}

The final step is $R_{22}^{-1} \cdot R_{21} \cdot R_{11}^{-1} \cdot R_{12}$

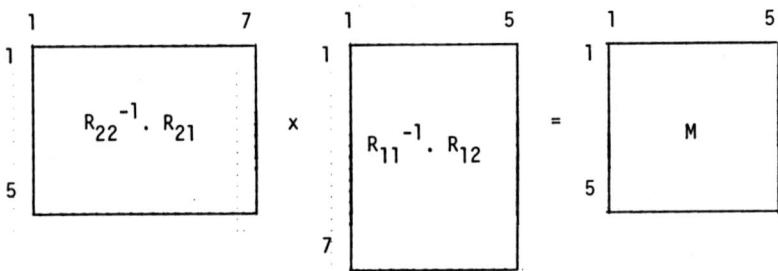

b) <u>The canonical roots.</u> In very crude terms, matrix M represents the ways in which the two sets of data covary with one another, and the latent roots of this matrix indicate the size of the common patterns. The occurrence of a single non-zero latent root indicates that, by an appropriate weighting of the criteria and the predictors, one set of variables can be projected into the space of the other along a straight line. The existence of two non-zero latent roots indicates that each set projects linearly into a two dimensional plane of the other set, and so on. The latent roots of matrix M are found satisfying

$$(M - \lambda I) = 0$$

Notice that by calling the smaller set of variables the predictors, we have simplified the problem of extracting the roots as the dimensions of matrix M are 5 x 5 and not 7 x 7. The numerics involved in the extraction of the roots can be followed in the worked example in section IV.

(iii) The significance of the roots

Each of the roots extracted from matrix M can be examined to see if it is statistically significant, or merely the product of chance factors. The statistic employed for this purpose is Bartlett's chi squared, which will be familiar to all students and practitioners of geography. In canonical analysis, as in most correlation studies, the null hypothesis is that the two sets of data are unrelated, and it is this assertion, rather than the research hypothesis, (that the two sets of data *are* related) that is tested by chi squared. The test is not applied to the data directly, but to a test statistic known as Wilk's Lambda (Λ) that is derived from the roots:

$$\Lambda = \prod_{i=1}^{p_2} (1 - \lambda_i) \qquad (\Pi \text{ means 'the product of'})$$

The null hypothesis, that the criteria are unrelated to the predictors, is tested by a function of Wilk's Lambda that is distributed approximately as chi squared, with pq degrees of freedom:

$$\chi^2 = - \left[(N-1) - 0.5 \ (p + q + 1) \right] \log_e \Lambda$$

If the null hypothesis can be rejected, and the two sets of data are found to be significantly related, the first latent root can be removed, and the significance of the remaining roots can be tested. In this way, chi squared provides a test of the significance of each root.

(iv) The canonical vectors

Having established by means of the latent roots, the size of the common patterns, we can now proceed to empirically define those patterns which are statistically significant. The meaning of each pattern is conveyed, firstly, by its two associated vectors of canonical weights, A and B, and secondly by its two associated vectors of canonical scores Sa and Sb. The canonical weights are by far the more difficult to estimate as our worked example will amply demonstrate! For our purposes, therefore it is sufficient to note that the weights B for the predictor variables are given by

$$(M - \lambda I) B = 0$$

and the weights A for the criteria variables by

$$A = (R_{11}^{-1} \cdot R_{12} \cdot B) \Big/ \sqrt{\lambda_1}$$

(v) The canonical scores

Once we have found the canonical weights, however, the estimation of the scores is easy. They are readily obtained by post-multiplying the standardised raw data by the appropriate sets of canonical weights. Thus if Zp denotes the standardised criteria variables, and Zq the standardised predictor variables, the scores Sa for the criteria are given by

$$Sa = Zp \ A$$

and the scores Sb for the predictors by

$$Sb = Zq \; B$$

IV A SIMPLE WORKED EXAMPLE

(i) The raw data

The steps in canonical correlation analysis can now be illustrated by reference to a simple worked example. For this purpose we take the smallest possible case, in which there are two variables in the criteria set, and two in the predictors, values of these variables being observed for six areas:

	Criteria p		Predictors q	
	CRIT 1	CRIT 2	PRED 1	PRED 2
ZONE 1	07	18	37	12
ZONE 2	11	13	18	40
ZONE 3	21	07	16	12
ZONE 4	20	15	09	11
ZONE 5	15	15	14	14
ZONE 6	11	01	01	11

(ii) Calculate means and standard deviations

MEAN ($\bar{x}$)	14.17	11.50	15.83	16.67
STANDARD DEVIATIONS (σ)	5.53	6.32	12.02	11.48

(iii) Standardise the raw data

Standardised data are required for the calculation of the canonical scores in step ix. (Most computer programs use standardised data for computing correlation co-efficients).

Standard scores, Z, are given by

$$Z = \frac{x - \bar{x}}{\sigma}$$

In our data:

	CRIT 1 Z_p	CRIT 2	PRED 1 Z_q	PRED 2
ZONE 1	-1.30	1.03	1.76	-0.41
ZONE 2	-0.57	0.24	0.18	2.03
ZONE 3	1.24	-0.71	0.01	-0.41
ZONE 4	1.06	0.55	-0.57	-0.49
ZONE 5	0.15	0.55	-0.15	-0.23
ZONE 6	-0.57	-1.66	-1.23	-0.49

(iv) <u>Calculate and partition the correlation matrix</u>

	CRIT 1	CRIT 2	PRED 1	PRED 2
CRIT 1	1.00	-0.15	-0.46	-0.28
	R_{11}		R_{12}	
CRIT 2	-0.15	1.00	0.70	0.15
PRED 1	-0.46	0.70	1.00	0.11
	R_{21}		R_{22}	
PRED 2	-0.28	0.15	0.11	1.00

(v) <u>Calculate the canonical roots</u>

To find the canonical roots, we first have to find the latent roots of the canonical equation:

$$(R_{22}^{-1} \cdot R_{21} \cdot R_{11}^{-1} \cdot R_{12} - \lambda I) B = 0$$

This operation can be broken down into several steps:

a) <u>Invert R_{22} and R_{11}</u>. Matrix inversing is normally rather complex, but since both R_{22} and R_{11} are symmetrical, their inverses can be found by first, reversing the signs of the elements in the non-principal diagonal and then, by dividing each element by a determinant d. The determinant is the product of the elements in the principal diagonal minus the products of the elements in the non-principal diagonal. So each element in R_{22} is divided by d = (1.00 x 1.00) - (-0.11 x -0.11) = 0.9879, and in R_{11}, d = (1.00 x 1.00) - (0.15 x 0.15) = 0.9775.

We have

$$R_{22}^{-1} \begin{pmatrix} 1.0122 & -0.1113 \\ -0.1113 & 1.0122 \end{pmatrix} \quad R_{11}^{-1} \begin{pmatrix} 1.0230 & 0.1534 \\ 0.1534 & 1.0230 \end{pmatrix}$$

b) <u>Multiply the sub-matrixes</u>. Form matrix M by multiplying the sub-matrixes

$$M = R_{22}^{-1} \cdot R_{21} \cdot R_{11}^{-1} \cdot R_{12}$$

$$\overset{R_{22}^{-1}}{\begin{pmatrix} 1.0122 & -0.1113 \\ -0.1113 & 1.0122 \end{pmatrix}} \quad \overset{R_{21}}{\begin{pmatrix} -0.46 & 0.70 \\ -0.28 & 0.15 \end{pmatrix}} \quad \overset{R_{11}^{-1}}{\begin{pmatrix} 1.0230 & 0.1534 \\ 0.1534 & 1.0230 \end{pmatrix}} \quad \overset{R_{12}}{\begin{pmatrix} -0.46 & 0.70 \\ -0.28 & 0.15 \end{pmatrix}}$$

$$R_{22}^{-1} \cdot R_{21} \qquad\qquad R_{11}^{-1} \cdot R_{12}$$

$$\begin{bmatrix} -0.4344 & 0.6919 \\ -0.2314 & 0.0739 \end{bmatrix} \cdot \begin{bmatrix} -0.3632 & -0.2634 \\ 0.6456 & 0.1105 \end{bmatrix}$$

$$\underbrace{\qquad\qquad}_{R_{22}^{-1} \cdot R_{21}} \cdot \underbrace{\qquad\qquad}_{R_{11}^{-1} \cdot R_{12}}$$

$$M = \begin{bmatrix} 0.6043 & 0.1908 \\ 0.1317 & 0.0690 \end{bmatrix}$$

c) **Extract the roots.**

The latent roots λ_1 and λ_2 are found such that by substituting either root in the following determinant yields zero:

$$\begin{bmatrix} 0.6043-\lambda_1 & 0.1908 \\ 0.1317 & 0.0609-\lambda_1 \end{bmatrix} = 0$$

As we are dealing in this example with only two variables, the relationships in the matrix form a simple quadratic function:-

$$(0.6043 - \lambda)(0.0690 - \lambda) - (0.1317 \times 0.1908) = 0$$

$$\lambda^2 - 0.6733\lambda - 0.0165 = 0$$

This can be solved using the familiar quadratic formula

$$\lambda = \frac{-b \pm \sqrt{b^2 - 4ac}}{2a}$$

$$\lambda = \frac{0.6733 \pm \sqrt{0.6733^2 - 4 \times 1 \times 0.0165}}{2 \times 1}$$

$$\lambda_1 = 0.6478$$
$$\lambda_2 = 0.0255$$

The canonical roots are the square roots of these values so

$$R \text{ canonical}_1 = 0.8044$$
$$R \text{ canonical}_2 = 0.1605$$

(vi) **Test the significance of the roots**

Two steps are involved in testing the significance of the latent roots:

a) **Define Wilk's Lambda**

$$\Lambda = \prod_{i=1}^{p_2} (1 - \lambda_i)$$

$$\Lambda = (1 - \lambda_1)(1 - \lambda_2)$$
$$\therefore \Lambda = (1 - 0.6478)(1 - 0.0255)$$
$$\Lambda = 0.3432$$

b) <u>Calculate Bartlett's chi - squared</u>

$$\chi^2 = -\left[(N - 1) - 0.5(p + q + 1)\right] \log_e \Lambda$$
$$\chi^2 = -\left[(6 - 1) - 0.5(2 + 2 + 1)\right] - 1.0700$$
$$\chi^2 = 2.6750$$

Degrees of freedom = pq = 4.

The chi squared value for the first latent root of 2.6750, with 4 degrees of freedom, is significant at the 0.6176 level. This means that it will occur six times out of ten by chance, so we are unable to reject the null hypothesis that the two sets of data are unrelated. In this, the smallest possible example of canonical correlation analysis, and using trivial data, the chi-squared test shows that there is no statistically significant relationship between the criteria and the predictors.

(vii) <u>Calculate the canonical weights for vector B_1</u>

The vector of canonical weights B_1 is found from the expression

$$(M - \lambda_1 I)B_1 = 0$$

In simple terms, the appropriate correlation sub-matrix, R_{22} is pre, and post multiplied by the cofactors V of any row of $(M-\lambda_1 I)$ so as to define a value θ which is the variance of the B_1 vector. The vector itself is given by

$$B = \frac{V}{\sqrt{\theta}}$$

In our example,
$$M = \begin{bmatrix} 0.6043 & 0.1908 \\ 0.1317 & 0.0690 \end{bmatrix}$$

and $\lambda_1 = 0.6478$

So substituting
$$\begin{bmatrix} 0.6043 - \lambda_1 & 0.1908 \\ 0.1317 & 0.0690 - \lambda_1 \end{bmatrix} = \begin{bmatrix} -0.0428 & 0.1908 \\ 0.1317 & -0.5781 \end{bmatrix}$$

The co-factors of this matrix are
$$\begin{bmatrix} -0.5781 & -0.1317 \\ 0.1908 & 0.0482 \end{bmatrix}$$

$$\theta = \text{first row of cofactors} \quad \times \quad \text{correlation submatrix } R_{22} \quad \times \quad \text{First row of cofactors}$$

$$\theta = \begin{bmatrix} -0.5781 & -0.1317 \end{bmatrix} \cdot \begin{bmatrix} 1.00 & 0.11 \\ 0.11 & 1.00 \end{bmatrix} \cdot \begin{bmatrix} -0.5781 \\ -0.1317 \end{bmatrix} = 0.3765$$

The weights b_1 and b_2 are given by

$$b_1 = \frac{V_1}{\sqrt{\theta}} = \frac{-0.5781}{\sqrt{0.3765}} = -0.942$$

$$b_2 = \frac{V_2}{\sqrt{\theta}} = \frac{-0.1317}{\sqrt{0.3765}} = -0.2146$$

(viii) <u>Calculate the canonical weights for vector A_1</u>

To get the corresponding vector of left hand weights for the first root we could go right back to the beginning and solve the canonical equation for A:

$$(R_{11}^{-1} \cdot R_{12} \cdot R_{22}^{-1} \cdot R_{21} - \lambda_1 I) A = 0$$

but a more convenient expression once we have reached this stage is:

$$A_1 = (R_{11}^{-1} \cdot R_{12} \cdot B_1) / \sqrt{\lambda_1}$$

So in our example:

$$A_1 = \begin{bmatrix} -0.3632 & -0.2634 \\ 0.6456 & 0.1105 \end{bmatrix} \cdot \begin{bmatrix} -0.9420 \\ -0.2146 \end{bmatrix} / \sqrt{0.6478} = \begin{bmatrix} 0.501 \\ -0.785 \end{bmatrix}$$

(ix) <u>Calculate the canonical scores</u>

The canonical scores are found by multiplying the standardised raw data by the appropriate set of canonical weights. The criteria scores for zone 1 on the first canonical pattern are given by

$$Sa = Zp \, A_1$$

$$Sa = \begin{matrix} \text{CRIT 1} \\ \\ \text{CRIT 2} \end{matrix} \begin{bmatrix} Zp \\ -1.30 \\ 1.03 \end{bmatrix} \cdot \begin{bmatrix} A_1 \\ 0.501 \\ -0.785 \end{bmatrix} = -1.4679$$

and the predictor scores for zone 1 on the first canonical pattern are given by

$$Sb = Z_q B_1$$

$$Sb = \begin{array}{c} \text{PRED 1} \\ \text{PRED 2} \end{array} \begin{bmatrix} Zq \\ 1.76 \\ -0.41 \end{bmatrix} \begin{bmatrix} B_1 \\ -0.942 \\ -0.214 \end{bmatrix} = -1.570$$

(x) <u>The complete results</u>

Steps (vii) to (ix) are repeated to find the canonical weights and canonical scores for the second pair of vectors A_2 and B_2.
The complete set of results is then:

Latent roots		Canonical roots	
λ_1 =	0.647	$R_{canonical_1}$ =	0.804
λ_2 =	0.025	$R_{canonical_2}$ =	0.160

<u>Canonical Weights.</u>

		A_1		B_1		A_2		B_2
Variables	CRIT 1	0.501	PRED 1	-0.942	CRIT 1	0.878	PRED 1	0.331
	CRIT 2	-0.795	PRED 2	-0.215	CRIT 2	0.624	PRED 2	-0.982

<u>Canonical Scores</u>

		Sa_1	Sb_1	Sa_2	Sb_2
	Zone 1	-1.467	-1.584	-0.496	0.981
	Zone 2	-0.475	-0.618	-0.354	-1.936
Areas	Zone 3	1.185	0.076	0.640	0.403
	Zone 4	0.087	0.648	1.127	0.296
	Zone 5	-0.365	0.196	0.478	0.177
	Zone 6	1.035	1.128	-1.153	0.076

(Note that there are some minor discrepancies between the values calculated in the worked example, and the set of complete results. The former were produced, to three decimal places on a hand calculator; the latter are the result of a computer analysis).

Despite the apparent complexity, the worked example demonstrates the basic relationships among the roots, vectors and scores. Where p=q=2, the analysis produces two canonical roots (neither of which, in this case, were found to be significant), two pairs of canonical vectors, and two pairs of canonical scores. All of these measures are somewhat tedious to derive even with the aid of an electronic calculator, and so, for all practical purposes (having first checked the results!) we can leave the number crunching to the computer. Several programs are now available; some departments will have access to the canonical analysis routine in the Kansas Geological Survey Series (Lee, 1969); alternatively, the canonical algorithm in the ICL Statistical Analysis Series (1972) is on file in most University and Polytechnic computer centres in Britain as is also Biomedical Computer Programs (B.M.D.). Having worked step by step through a simple example, however, we can now leave the mathematics behind. For our final treatment of the technique, we turn to the results of an actual geographical application.

V A GEOGRAPHICAL APPLICATION

(i) Theory and data

In this section, we are concerned with empirical applications of canonical correlation analysis. For this purpose it is most useful to report, in some detail, a specific geographical study. Our example involves an important theoretical statement in human geography, B.J.L. Berry's general field theory of spatial behaviour (1966). The field theory suggests that there is an isomorphic (i.e., a structurally equivalent) association between land using activities and patterns of movement through space. For example, a city cannot exist in isolation, it is dependent for its maintenance upon the continual inflow and export of energy in the form of commodities, personnel and ideas. Similarly, these spatial flows are determined by, and in turn make possible, the system of urban centres. Activities and movements combine to produce identical structures in space, their relationship is one of interdependency.

It was via the field theory that Berry (1966) first introduced canonical correlation to geography, in a study of spatial structures in India. More recently, the associations between formal and functional patterns were examined in an analysis of the regional structure of Wales (Clark, 1973a). The data consisted of the observed values of five activity and eight linkage dimensions for a set of seventy-two areas in Wales. The first set of variables, the criteria, measured, respectively, the urban, tourist, agricultural, mining, and coastal industrial importance of each of the areas, while the eight linkage variables (the predictors), specified the involvement of each of the zones in the dominant patterns of information flow in the region. These variables were named after the major focal centres within each flow pattern. Both sets of variables were, in fact, factor outputs from a principal components analysis of raw data, so that the variables satisfied most of the distributional requirements for the canonical model (see section VI, i). The field theory asserts the interdependency of movement and activities so that the problem, in simple terms, was to identify the extent and nature of the overlap between these two separate measures of spatial structure.

The associations were specified by bringing the two sets of data together in a canonical framework. If the patterns of formal and functional structure have a simple one - to - one relationship, the canonical analysis should pick off pairs of vectors from the 13 x 13 intercorrelation matrix. Five of the eight linkage patterns will be related to the five activity patterns, and three of the linkage patterns will fail to weight significantly on any of the vectors. If, however, the activities and linkages overlap in more complex ways, then new vectors of canonical weights will be created on either side expressing these interrelationships. Finally, there may be little or no similarity in the structure of space as it is defined by the activity and the linkage dimensions. In this case, the canonical analysis would fail to produce a single significant root, reflecting the failure of the vectors to integrate meaningfully the two sets of data.

(ii) Results

The canonical analysis identified the major associations between the two independent measures of spatial structure. It indicated both the size of the common patterns, and the variables and areas which were involved in them, (though the canonical scores were not reported in the original paper). The levels of significance of the five canonical roots pointed to the existence of a number of dominant and subdominant relationships in the structure of Wales (Table 5). The first three canonical correlation coefficients were highly significant, though the remaining two roots, and the fifth in particular were of lesser importance. The empirical meaning of the common patterns was expressed by the canonical weights, and these were interpreted so as to show the ways in which the area 'hangs together'.

Table 5 The Structure of Wales: Canonical Roots

Number of Roots Removed	Largest Root Remaining	Corresponding Canonical Correlation	Wilk's Lambda	Bartlett's Chi Square	Degrees of Freedom	Probability
0	0.901	0.949	0.003	354.249	40	0.000
1	0.808	0.899	0.040	205.698	28	0.000
2	0.697	0.835	0.210	99.844	18	0.000
3	0.216	0.465	0.694	23.342	10	0.020
4	0.113	0.337	0.886	7.724	4	0.210

Table 6 The Structure of Wales: Canonical Vectors

	\	\	Vector	\	\
	One	Two	Three	Four	Five
Structural variables					
1 Urban status	<u>0.870</u>	0.474	0.036	0.124	-0.033
2 Holiday industry	0.470	<u>-0.801</u>	0.348	0.082	0.092
3 Agriculture	0.063	0.146	0.055	<u>0.932</u>	<u>0.820</u>
4 Mining communities	0.117	-0.238	0.577	0.244	<u>-0.732</u>
5 Coastal activity	0.080	-0.225	<u>0.739</u>	0.212	0.591
			Vector		
Linkage variables	One	Two	Three	Four	Five
I Cardiff	<u>0.811</u>	0.610	-0.089	0.147	-0.030
II Colwyn Bay	<u>0.562</u>	-0.417	-0.297	0.096	-0.138
III Swansea	<u>0.310</u>	-0.267	<u>0.620</u>	0.018	-0.048
IV Shrewsbury	<u>0.219</u>	0.237	0.090	<u>0.809</u>	0.303
V Newport/Pontypool	<u>0.287</u>	0.027	0.565	0.211	<u>-0.694</u>
VI Aberystwyth	0.056	-0.041	0.369	0.281	<u>0.842</u>
VII Chester	<u>0.298</u>	0.095	0.151	0.177	0.134
VIII Rhyl	<u>0.377</u>	-0.555	0.055	0.276	0.285

_____ Leading weights are underlined.

(iii) Interpretation

The vectors were interpreted by considering the magnitude of their weights in relation to the direction of their signs. In view of the highly complex nature of the settlement pattern in Wales, it was not surprising to find that the canonical analysis combined the two sets of variables in a number of different ways (Table 6). At first glance, the first pair of vectors appears simply to emphasise the relationship between the concentration of linkages upon Cardiff, and its importance as a major centre of service and quaternary activity in Wales. A closer examination, however, suggests that the vector goes further than a mere concentration upon the capital city since, with the exception of Aberystwyth (VI), the weighting of all the linkage variables is high positive. This indicates that the basic relationship in the spatial structure of Wales is the concentration of connections upon those centres which have the highest urban status. In this way, vector one confirms the importance of a well established and much researched relationship in urban geography; the framework of a settlement system consists of a set of primary flows determined by, and supporting, the central place activities of the area.

A second overlap of parameters is revealed by the weights on canonical vector two by combining one of the activity with two of the linkage dimensions. The nature of this relationship is fairly simple since it is based upon those places in the region whose patterns of associations of linkage correspond to their level of importance as centres of holiday industry. The vectors indicate that places important for tourism (Rhyl, Colwyn Bay, and to a lesser extent, Swansea) are important also as nodal centres and as origins feeding connections to the holiday areas. The spatial structure of Wales is therefore integrated in this second way by a set of flows responsible for, and resulting from, the distribution of holiday activities in the area.

Vector three highlights a structural relationship in South Wales, as the leading weights are associated with the Swansea and Newport/Pontypool subsystems, and with the coastal activity variable. The pattern of flows focusing upon Swansea and Newport/Pontypool is such that, when taken together, they overlap so as to define a complex system of interlinkage over the whole of South Wales. Against this background, the empirical meaning of vector three becomes clear as it involves those areas in South Wales which exchange connections with a set of coastally located service and industrial activities. In terms of the economic geography of the area, it could be termed a 'hinterland vector' since it involves a set of flows arising out of, and maintaining, the urban/industrial activities of the south coastal fringe.

Two less important sets of relationships were additionally specified by the canonical analysis, indicating further ways in which Wales could be formally and functionally integrated. Vector four, and the positive side of bipolar vector five, underlines an isomorphism within the system between sets of rural connections and occupations. Of the five interrelationships specified, this is perhaps the most easy to conceptualise, since it involves no more than a visual superimposition of two flow systems, those of Shrewsbury and Aberystwyth, upon those areas shown by the structural variables to be important for agricultural activity.

The remaining interdependency revealed by the canonical analysis was specified by the negative pole of vector five. By combining the fourth structural with the fifth linkage variable it suggests that the flows which focus upon Newport and Pontypool arise from, and maintain, the mining communities of eastern South Wales.

Having discussed each interrelationship in turn it is now possible to specify how the formal and functional elements combine. Three dominant and two subdominant interdependencies are defined and these in order of importance are:

a) A general exchange of central place connections involving all areas according to the size and range of their urban functions.
b) A predominantly north, and north east coastal concentration of linkage

associated with and maintaining places which are important for their tourist activities.
c) An intra South Wales pattern flows associated with the industrial centres of the coastal belt.
d) An intra rural exchange of connections in the north east and the south west associated with the agricultural activities of these areas.
e) A localised interchange among the mining communities of the central and eastern regions of the South Wales coalfield.

(iv) Evaluation

The five canonical vectors summarise the main ways in which the formal and functional patterns combine. In this manner, canonical correlation analysis provides a general test of the field theory as well as identifying the major components of regional structure. The primary value of this example for our purposes, however, is that it illustrates some of the advantages as well as some of the difficulties that arise in canonical applications in geography. A major advantage, seen in the summary descriptions of the vectors, is that it leads to comparatively simple statements being made about complex spatial patterns, while a further asset is that as the patterns are statistically independent, they are capable of further analysis and testing. The major problem is clearly that of naming the vectors for, with the possible exception of vector one, none of the patterns was particularly easy to disentangle. Moreover, decisions as to which weights to select is largely a matter of personal choice. What about the positive weights on vector two? Why not consider variable 4 in the description of vector three? In the absence of any meaningful tests of statistical significance for the weights, these very pertinent questions cannot be answered; the final interpretation remains (as it does in most other forms of multivariate analysis), a matter of reasoned but, nevertheless, subjective judgement. While there are, however, underlying technical reasons for the difficulties of interpretation (see section VI, iv), the problems in this instance (as in the studies of Openshaw, 1969, and Ray, 1971), are as much a product of the overall research design as an inherent feature of the technique. Given the level of analysis, it would be unreasonable to expect complex spatial variables to interlock in a straightforward way: the real world as all geographers know, is not that simple!

A second point to stress is that the present example illustrates merely *one* of the ways in which canonical correlation analysis may be used in geography. In 1966, Kendall and Stuart claimed that canonical correlation "is best regarded as an exploratory tool which will give us some idea of the structure of the multivariate complex under study" (p.305). This use of canonical analysis as a technique of preliminary investigation is represented in geography by the work of Norcliffe (1972), and Gauthier (1968). The Welsh study is, however, similar to several others (notably those of Berry, 1966; Andrews, 1973; Openshaw, 1969; and Ray, 1971) in that canonical correlation is used as the *final*, rather than the initial stage, in the analysis. Its role in this context is to draw together and integrate those aspects of the spatial system which previous analysis had considered in isolation and apart. Thus both the activity and the linkage variables were first analysed separately before being considered together. Canonical correlation analysis provides the analytical format within which this process of synthesis can be achieved. Such a facility is invaluable from a purely technical point of view since there is a deficiency of high level measures of multivariate association in statistics. In a wider methodological context it serves to facilitate general statements concerning the system as a whole, rather than unique descriptions of unique elements.

VI PROBLEMS AND POSSIBILITIES

Despite the growing popularity of canonical correlation analysis in geography, the level of technical evaluation and comment remain low. There is a major contrast in this respect with geographical procedures such as trend surface analysis, and factor/component analysis, which are supported by an extensive and detailed literature dealing with problems of analysis and application. (This literature is reviewed in *The Statistician* 23, 1974, which is devoted entirely to the use of statistics in geography). If the potentials of canonical correlation are, however, to be fully realised, then the major problems surrounding the use of the technique require careful evaluation. Equally, we have discussed merely the simplest form of canonical correlation, and the model is capable of certain extensions, some of which make it more useful for geographical purposes. This monograph therefore concludes by briefly listing the problems and possibilities under four broad headings.

(i) The characteristics of the input data

Canonical correlation, in common with most multivariate procedures, is a parametric technique. One feature of parametric models is that they impose certain constraints upon their input data. The main assumptions are that the variables are normally distributed, are without measurement error, are linear, homoscedastic, and serially independent (Poole and O'Farrell, 1971). These are strict theoretical requirements but are rarely encountered in empirical reality: indeed the failure of the data to satisfy all the *relevant* requirements is a feature of most operational designs. But in spite of this situation it is difficult to offer any clear advice as to procedure. When do data become non-normal? When are they heteroscedastic? At what stage does the failure to meet the requirements become critical to the performance of the model? In general, these questions have no precise answers; the success of canonical correlation analysis, indeed of any statistical procedure is dependent, in the final analysis, upon 'reasonable' and 'sensible' application.

Two data problems however merit more specific comment. One is the normality assumption which is critical at two stages in the analysis. The first occurs when the similarities among the variables are summarised in the form of product moment correlation coefficients, which are themselves parametric measures. The common patterns within this matrix are then expressed by the canonical weights. The second critical step is when the raw data are standardised (see section IV, iii) prior to being post-multiplied by the weights so as to form the canonical scores (IV,ix). The scores convey the spatial information in the data, and yet are doubly sensitive to departures from normality. As in principal components analysis, they are derived by second stage description from two matrices, each of which makes assumptions of normality in the data (Clark, 1973c).

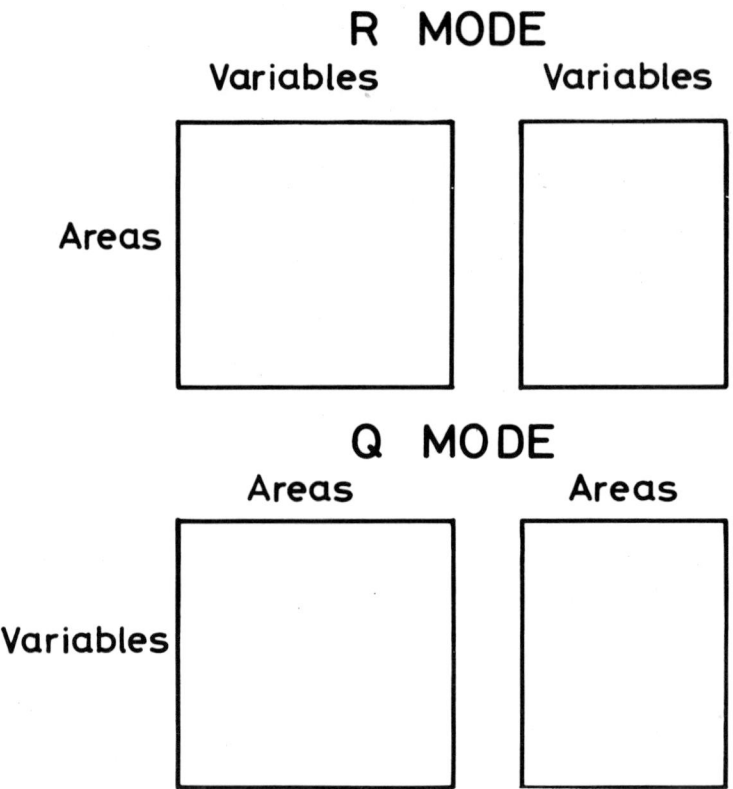

Fig. 3 R & Q mode data formats for canonical correlation analysis. In R-mode analysis the weights relate to the variables and the scores to the areas: in Q mode, the weights refer to the areas and the scores to the variables.

The second data problem is that of multicollinearity, for canonical correlation analysis appears to be highly sensitive to the presence of intercorrelation within the data. Multiple regression analysis is affected by correlation within the independent set, but canonical correlation, with two sets of data, suffers from a double disadvantage in that there may be a high level of intercorrelation within either or both sets. In both techniques, the analysis involves the inversion of correlation matrices which, when multicollinearity is present, may be non-singular. This, according to Gittus and Stephens (1974) leads to unintelligible results.

(ii) Data formats

A feature of most geographical applications of canonical correlation analysis is the use of the R-mode data format. In R-mode analysis, the variables (the columns) in the matrix, are correlated over the areas (the rows) so that the weights denote the involvement of each variable in each of the common patterns, and the scores indicate the involvement of each of the areas. Given the geographers traditional concern with areas and regions, this format is somewhat paradoxical since one would have expected more attention to be focused upon Q-mode analysis in which the areas are correlated over the variables (fig. 3). In this situation, the labels are reversed with the weights applying to the areas and the scores to the variables. One of the many advantages with the Q-mode format is that the double normality problem influences the variables (the scores) rather than the areas and so leads to the identification of more meaningful common spatial patterns.

(iii) N - Modal canonical analysis

Another possible development is to incorporate additional sets of data into the analysis. Although the standard canonical model as outlined by Hotelling (1936, 1937) is concerned with the relationship between two data sets, the technique can in fact be extended to n sets. Such a model might be used to identify the common elements in the employment and spatial structure of a region over *several* points in time (i.e. censuses) rather than merely two, as in our example in section II. Needless to say, problems of data inputs, and of interpretation would increase exponentially in such applications!

(iv) Interpreting the results

One consequence of the comparatively few canonical studies reported in the geographical literature is that there is no overall concensus as to procedure in interpretation. This applies particularly to the table of scores, potentially the most valuable for the geographer since it relates, (in R - mode studies) to areas, but one which has been almost totally ignored. Norcliffe (1972), and Green, Halbert and Robinson (1966), in simple teaching examples, are two of the very few to report and comment on the score values, but there is no mention of them in studies by Berry (1966), Clark(1973a, 1973b), Gauthier (1966), Openshaw (1969) or Ray (1971). The scores denote the degree and direction of involvement of the observed values of the criteria and predictor variables in each of the common patterns. When the latent root is 1.00, indicating perfect correspondence, the values in each pair of scores is the same, and, as the latent root reduces in size, the values diverge, (Table 4). One means of interpretation is, therefore, to isolate graphically, those score values that deviate most from the line of perfect correlation (fig. 4). As canonical analysis involves no concepts of dependence or independence, it is not possible to say why individual values deviate. All we can say is that we have a common spatial pattern in which some areas (the non-deviants) are centrally involved, and others (the deviants), marginally involved. The geographical meaning of this pattern can be expressed by mapping each of the values according to the magnitude of its deviation.

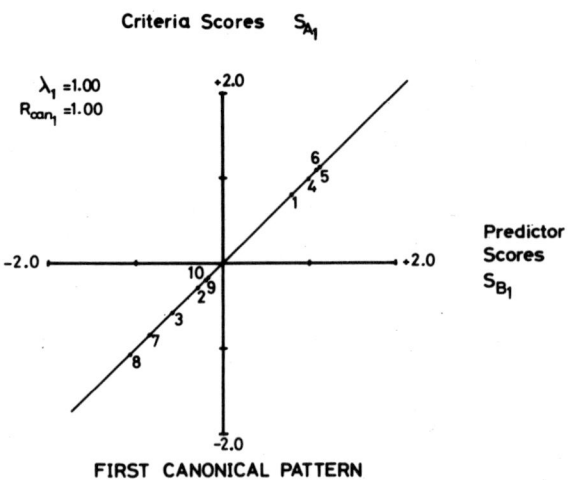
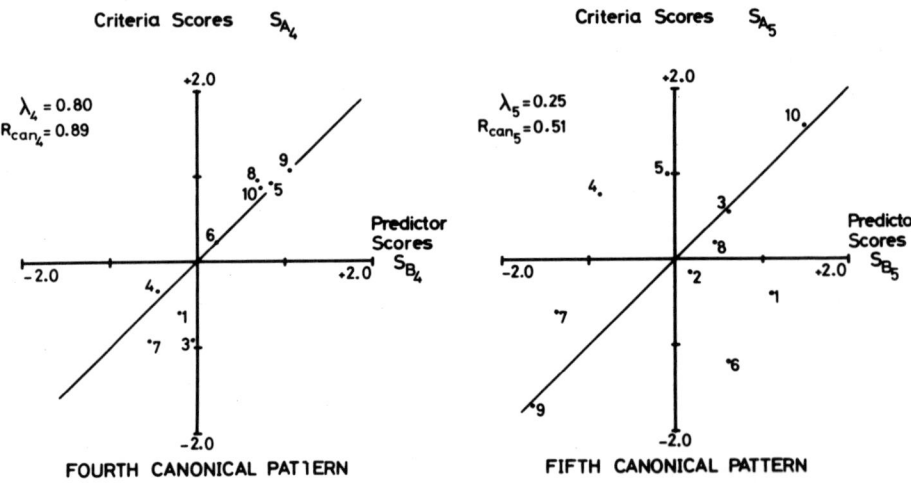

Fig. 4 Hypothetical employment data: plots of selected vectors of canonical scores

Earlier, in Section I, we referred to Kendall and Stuart's beliefs (in 1966), "that the results of canonical correlation analysis are more difficult to interpret than those of factor analysis". One reason, according to Monmonier and Finn (1973) lies in the nature of the weights. While the weights are somewhat akin to factor loadings in that they represent the relative contribution of the original variables to the canonical vector, they are not, as in principal component loadings, correlations between the vectors and the original variables. (The indirect nature of the relationship between the variables and the weights is shown in the worked example, section IV, vii). These correlations (canonical loadings) can, however, be computed so as to improve the intelligibility of the scores, by pre-multiplying the matrix of canonical vectors by the partitional intercorrelation matrix (Monmonier & Finn, 1973); Gittus & Stephens, 1974). But despite this enhancement it must be admitted that the interpretation issue, in common with all the other problems and possibilities of canonical correlation analysis, requires considerable further study.

VII CONCLUSION

This monograph began by stressing the complexities of canonical correlation, and concluded by highlighting some of its more obvious deficiencies. It would be unfortunate if this problem - oriented approach were to deter the reader from using the technique since no science, and least of all geography, can afford to ignore those procedures which might lead to the production of high level generalisations. Indeed, the growing number of geographical applications reflects a general recognition of its values in this respect. But these are clearly very early days. Principal components analysis was first introduced to geography in 1939, though it did not become a standard analytical tool until the early Sixties. Assuming even a much reduced time scale dating from 1966, it is clear that the full potentials of canonical correlation analysis have yet to be realised.

BIBLIOGRAPHY

A. BASIC TEXTS FOR MATHEMATICAL TREATMENT

Anderson, T.W. (1958), *An introduction to multivariate statistical analysis.* Wiley, New York.

Cooley, W.W. & Lohnes, R.P. (1971), *Multivariate data analysis.* Wiley, London.

Hope, K. (1968), *Methods of multivariate analysis.* University of London Press

Kendall, M.G. (1957), *A course in multivariate analysis.* Griffin, London.

B. STATISTICAL TESTS RELATING TO CANONICAL ANALYSIS

Bartlett, M.S. (1941), The statistical significance of canonical correlations. *Biometrica,* 32, 29-38.

Lawley, D. N. (1959), Tests of significance in canonical analysis. *Biometrica,* 46, 59-66.

C. COMPUTER PROGRAMS FOR CANONICAL CORRELATION ANALYSIS

B.M.D., (1965), *Biomedical computer programs.* Health Science Computing Facility, U.C.L.A. School of Medicine. Los Angeles.

I.C.L., (1972), *Statistical Analysis Mark 3.* (XDS.3).

S.P.S.S., (1972), *Statistical package for the social sciences.* National Opinion Research Centre, Chicago.

Lee, J.P. (1969), FORTRAN IV Programmes for canonical correlation and canonical trend-surface analysis. *Computer Contributions: Kansas Geological Survey,* 32.

D. CANONICAL APPLICATIONS AND EXAMPLES

Andrews, H.F. (1973), Urban structure correlates of tertiary activity. *Regional Studies,* 7, 263-70.

Barnett, V.D. & Lewis, T. (1963), A study of the relations between G.C.E. and degree results. *Journal of the Royal Statistical Society,* A,126, 187-95.

Berry, B.J.L. (1966), Essays on commodity flow and the spatial structure of the Indian economy. *University of Chicago, Department of Geography, Research Paper 111.*

Clark, D. (1973a), The formal and functional structure of Wales. *Annals, Association of American Geographers,* 63, 71-84.

Clark, D. (1973b), Urban linkage and regional structure in Wales: in analysis of change. 1958-1968. *Transactions, Institute of British Geographers,* 58, 41-58.

Gauthier, H.L. (1968), Transporation and the growth of the Sao Paulo economy. *Journal of Regional Science,* 8, 77-94.

Gittus, E. & Stephens, C.J. (1974), Some problems in the use of canonical analysis; with an example from urban sociology. (Mimeo).

Gould, P. (1968), How to shoot the canonical correlation. University of Ohio, (Mimeo.).

Green, P.E., Halbert, M.H. & Robinson, P.J. (1966), Canonical analysis: an exposition and illustrative application. *Journal of Marketing Research,* 3, 32-9.

Monmonier, M.S. & Finn, F.E. (1973), Improving the interpretation of geographical canonical correlation models. *Professional Geographer,* 25, 140-2.

Norcliffe, G.B. (1972), Canonical analysis of the relations between certain aspects of the demographic and urban systems of the Republic of Ireland. *Irish Geography,* 6, 411-427.

Openshaw, S. (1969), Canonical correlates of social structure and urban building fabric. *University of Newcastle-upon-Tyne, Department of Geography, Seminar Paper* 11.

Ray, D.M. (1971), From factorial to canonical ecology, the spatial interrelationships of economic and cultural differences in Canada. *Economic Geography,* 47, 344-55.

Willis, K.G. (1972), The influence of spatial structure and socio-economic factors on migration rates. *Regional Studies,* 6, 69-82.

E. SUPPLEMENTARY READING

Clark, D. (1973c), Normality, transformation and the principal components solution; an empirical note. *Area,* 5, 110-113.

Cole, J.P. & King, C.A.M. (1968), *Quantitative Geography,* Wiley, London.

Gregory, S. (1963), *Statistical methods and the geographer.* Longmans, London.

Hotelling, H. (1936), Relations between two sets of variates. *Biometrica,* 28, 321-9.

Hotelling, H. (1937), The most predictable criterion. *Journal of Educational Psychology,* 26, 139-42.

Kendall, M.G. & Stuart, A. (1966), *The advanced theory of statistics.* Griffin, London.

King, L.J. (1969), *Statistical analysis in geography*. Prentice Hall. Englewood Cliffs N.J.

Meredith, W. (1964), Canonical correlations with fallible data. *Psychometrica,* 29, 55-65.

Poole, M.A. & O'Farrell, P.N. (1971), The assumptions of the linear regression model. *Transactions, Institute of British Geographers,* 52, 145-58.

Theakstone, W.H. & Harrison, C. (1970), *The analysis of geographical data,* Heinemann, London.